IMAGES
of America

EAGLE ROCK
RESERVATION

This wood engraving depicting Eagle Rock was published by D. Appleton & Company in 1873. It was included in William Cullen Bryant's subscription series *Picturesque America*. The wood engravings from this series are in many cases more detailed and dramatic than the steel engravings that were more highly prized at the time of publication. These fine prints represent the skill of the wood engraver and are examples of top-quality detailing prior to other illustration methods becoming popular. This engraving shows the eagles that once nested among the crags outlining the rugged contour of the mountainside. From this, Eagle Rock derives its name. The artist possibly drew a picture of Eagle Rock to be used in one of the many period postcards showing a similar view. Today, this hillside below Eagle Rock is overgrown with many trees along Mountain Avenue on the West Orange–Montclair border.

IMAGES
of America

EAGLE ROCK
RESERVATION

Joseph Fagan

ARCADIA
PUBLISHING

Published by Arcadia Publishing
Charleston, South Carolina

Printed in the United States of America

Library of Congress Catalog Card Number: 2002110811

For all general information contact Arcadia Publishing at:
Telephone 843-853-2070
Fax 843-853-0044
E-mail sales@arcadiapublishing.com
For customer service and orders:
Toll-Free 1-888-313-2665

Visit us on the Internet at www.arcadiapublishing.com

*This book is dedicated to Anthony Olef (1636–1723),
considered to be the first settler near Eagle Rock. No street bears his
name; no monument marks his existence. The spirit of perseverance
that he undoubtedly brought with him to the wilderness then still
survives today in a tract of land called the Eagle Rock Reservation.*

Shown here is the cover of an advertising brochure used to promote the resort at Eagle Rock
c. 1895. Eagle Rock is located in West Orange, Essex County, New Jersey, in the Orange
Mountains, part of the Watchung Mountain Range.

CONTENTS

ACKNOWLEDGMENTS

I would like to express my sincere appreciation to the following people and organizations, in no particular order, who have helped make this book possible: the West Orange Public Library, the Montclair Public Library, the Newark Public Library, the New Jersey Historical Society, the Essex County Department of Parks and Cultural Affairs, the Knowles family, the Maxwell family, Rosemarie and Cliff Platt, and Kenneth Tinquist. Thanks also go to my research assistants—my wife, Debbie, and our son Joseph—and to the countless friends and family members too numerous to list for all their support and encouragement. Finally, I feel a deep debt of gratitude to my parents, James and Helen Fagan, whose love and patience taught me the confidence and courage to believe in myself.

INTRODUCTION

Eagle Rock will not appear on any list of national historic places. No great battle ever took place there, and no significant event of American history occurred within its boundaries. However, someday the glaciers that forged out the valley below during the last ice age will return and perhaps pause in amazement over the spectacular vista carved on their previous visit. Eagle Rock is abound with natural beauty—mostly unchanged over the centuries—and stands as an undisturbed link to a largely forgotten past. The 20th century has grown up all around it, but the park still exists as if an island in a vast sea of urban sprawl. When the trolley line was torn up and the resort hotels closed down, no one seemed to notice or care, and the reservation slowly faded from memory. Yet time has been kind to Eagle Rock, and its appeal, as if a treasured heirloom, has been passed from one generation to the next. Eagle Rock enters the 21st century poised and proud to once again be discovered and enjoyed. It survives today as a timeless monument for all who can see the inherent splendid beauty. This book tells its story.

My parents first took me to the scenic overlook on July 4, 1959, to view fireworks. Even as a youngster, something about it caught my interest. I grew up in the shadow of Eagle Rock and throughout my childhood walked its trails and paths with my family and friends. Like my father before me, I considered the park my personal playground. My father was no stranger to Eagle Rock, having grown up in West Orange himself. The tree in which he once carved his initials in the 1930s still stood (up to just a few years ago) along the winding road. He often spoke of the old Indian trail where he played as a child. It was very close to Eagle Rock and ran along the steep nearby cliff. He and his friends used to swing out over the cliff on a rope securely tied to a hanging tree limb. I suppose knowing how my father played in Eagle Rock as a kid is what sparked my own curiosity at a young age. Even my grandparents spent lazy Sunday afternoons picnicking in the park with the family. It is easy to see why, for me, Eagle Rock has become more then a place, it has become a loyal friend, always there greeting me with open arms. My son and his cousins, as far away as Virginia and Maryland, are fourth-generation Fagans who know our dear family friend: Eagle Rock.

So many people have visited Eagle Rock over time that our family's connection with it is not unique. Many have returned to these enchanting grounds to retrace their past with their children and even grandchildren. Eagle Rock is special because it remains timeless and full of a rich history so often and easily overlooked. Although part of the Essex County Park System, its location in West Orange and neighboring towns has permitted it to become interwoven into the fabric of many communities. People come from near and far to enjoy it as a refreshing distraction from the sometimes hectic routine of daily activities. It is best known for the panoramic view of the New York City skyline, which can be breathtaking on a clear day. There are also miles of trails and paths to explore its natural and diverse beauty.

Often overlooked is the history of Eagle Rock. Before it became part of the then newly formed Essex County Park Commission in 1895, it was actually planned to be part of Llewellyn Park. In 1854, Llewellyn Haskell began building his house, named "the Eyrie," on the foundation of an old farmhouse he had purchased from Henry Walker in 1853. Haskell admired nature and fell in love with the abundance of natural beauty at Eagle Rock and the entire first mountain. He voiced ambitions to build a boulevard along the mountain ridge, 100 feet in width, with his residence serving as the entrance. The dream was never realized, but from it spawned the idea of nearby world-renowned Llewellyn Park, which bears his name and survives today. The Eyrie became a landmark at Eagle Rock and survived until 1924, when it was finally torn down.

Most would be surprised to learn that before the start of the 20th century, Eagle Rock was considered a resort. On the open field once sat a hotel, a cottage, a cafe, a beer garden, and in later years, an ice-cream pavilion. No automobiles were around to take one to Eagle Rock. Other than by walking, access could be gained by trolley. In fact, Eagle Rock amazingly had a trolley four years before the neighboring town of Montclair because of its popularity.

Thomas Edison's laboratories stood nearby in West Orange, but he also conducted secret experiments for the U.S. Navy during World War I at Eagle Rock. In the Casino, a building constructed in 1909, he performed nearly 103 experiments. Many of these still remained classified by the navy until just a few years ago. That building was also a landmark at Eagle Rock for years but fell into a state of disrepair. In the 1980s, the Casino was saved from destruction and was renovated; it survives today as Eagle Rock's new landmark, the Highlawn Pavilion.

Since great effort has been taken in describing the present-day location of many of these views, this book can be used as a field guide. Take it to Eagle Rock while visiting and discover its forgotten history. Stare across the open field and try to imagine the sights and sounds of yesteryear. Through the silence, envision the once romantic appeal of the Eagle Rock of a century ago.

One

Llewellyn Haskell and the Early Years

This wood engraving from the *New York Illustrated News*, June 23, 1860, is perhaps the earliest depiction of Eagle Rock. This view looks northeast from the cliff at Eagle Rock. In the distance is the Hudson River, and in the left foreground is the present location of Montclair, then known as West Bloomfield.

This 1895 drawing from *Essex County, NJ Illustrated* shows the first surveying station established at Eagle Rock by the first settlers in Newark shortly after their arrival in 1666. No one could have ever dreamed that in less then three centuries this vast area would be inhabited by more than 12 million people.

During the Revolutionary War and most likely *c.* 1780, the Continental Army, under the command of George Washington, used the vantage point from Eagle Rock as one of a chain of observation posts extending from Paterson to Summit. From here, it was possible to monitor British troop movements in and around New York and to protect the countryside from Tory raiders.

Appearing in an 1884 edition of *Harper's Weekly*, this drawing depicts the cliffs just a short distance from Eagle Rock. An old Indian trail ran along the top of this mountain, and the area operated as a quarry until the early 1980s. The drawing shows the scenic wonder and beauty of the many rock formations existing in and around Eagle Rock.

This architectural drawing by famous architect Alexander J. Davis appeared in the *New York Illustrated News* in 1860. Depicted here is the home of Llewellyn Haskell, who first purchased land at Eagle Rock in 1853. This house, designed by Davis, was named the Eyrie. It became an Eagle Rock landmark and retained the same name until it was torn down in 1924.

This studio picture of Llewellyn S. Haskell (sitting) was taken by J. Kirk Photographers on Broad Street in Newark, New Jersey, c. 1865. The women remain unidentified but are perhaps his wife and daughter. Haskell made his first land purchase at Eagle Rock from Henry Walker in 1854 and built his house, the Eyrie, there. The house was constructed on the foundation of an old farmhouse. His love of nature and the beauty of Eagle Rock inspired his vision of preserving land for future generations. He founded Llewellyn Park in West Orange, which was America's first planned community. It survives, mostly unchanged from its humble beginnings more than a century ago. Eagle Rock was originally planned to be part of Llewellyn Park, but this never happened. In 1895, the newly formed Essex County Park Commission began to purchase land at Eagle Rock. Haskell died in 1872 and perhaps never realized the impact of his lasting legacy.

The two women pictured here at the Eyrie *c.* 1870 are unidentified. The man seems to be Llewellyn Haskell. "Eyrie" means a nest of a bird of prey. This name was given to the house presumably in reference to the eagles that once occupied Eagle Rock.

This *c.* 1857 map of Llewellyn Park shows the Eyrie at Eagle Rock as part of Llewellyn Park. It is depicted as the focal point of the map on the featured vignette. Also shown is an observation tower at the edge of the cliff on Eagle Rock built by Haskell. This tower toppled in a violent storm; Haskell narrowly escaped with his life.

The location of the Eyrie is somewhat recognizable in this *c.* 1908 photograph, for the lay of the land has not changed all that much. The house stood at the eastern end of what is today the upper picnic area. The small opening in the trees in the lower right still leads down a path to a pay phone at Eagle Rock.

The wisdom of building the circular-shaped dwelling was questioned by many of the townspeople, but the operations proceeded. Pieces of bark were removed from trees specially selected by Haskell and fastened to the framework of the house. The circular tower was made of traprock quarried from the nearby mountain.

By July 9, 1916, the Eyrie had passed through several hands. The building, which always retained its original name, had its share of troubles. The last private owner, Cardelia A. Graham, purchased it by foreclosure on October 9, 1897, from the Essex County sheriff, Henry Doremus, who later became mayor of Newark. In January 1898, it was purchased by the then newly formed Essex County Park Commission for the price of $14,903.36. When this photograph was taken, Haskell's former home was nothing more than a curious old house badly in need of repair. The last few owners prior to the Graham family had not even lived in the house year-round. Haskell had owned the house from 1854 until 1871 but most likely had not lived at the Eyrie in the later years. (Courtesy of the Newark Public Library.)

The Eyrie was last photographed on February 12, 1924, just prior to being torn down. As early as 1904, however, it had begun deteriorating and decaying. Over the years, it was battered by winds and storms, and the roof had partially fallen in and was in danger of total collapse.

A closer look clearly shows the house's state of disrepair. A combination of broken glass, boarded-up windows, and rotting bark meant only a total renovation could save it, and the funds for such a venture did not exist. By this time, the once famous home was nothing more than a decaying monument to a prestigious past.

This *c.* 1870 photograph depicts not the Eyrie but the gatehouse at the main entrance to Llewellyn Park in West Orange, near Park Avenue. Built *c.* 1860, the gatehouse is a replica of the Eyrie. Dietrich Everett, the stonemason on Haskell's house at Eagle Rock, was the first gatekeeper here. The house stands today; only the surrounding landscape has changed.

LLEWELLYN PARK,
ORANGE, NEW JERSEY.

MAIN GATEWAY AND LODGE OF LLEWELLYN PARK

Llewellyn Park, at Orange, N. J., combines acknowledged healthfulness, accessibility, and social advantages, with an opportunity to secure the enjoyment of a large and costly country place, by the purchase of one acre of land, which carries with it the possession of fifty acres of pleasure-grounds, and seven miles of private drives, throughout a tract of land nearly as large as Central Park. All persons seeking a country home are invited to visit and examine the remaining building-sites of from one to five acres each, the greater portion of the Park having already been sold, and improved by New York Merchants.

For Maps, Terms, and Particulars, apply to the owner, **L. S. HASKELL, Orange, N. J.,**
Or to **HAZARD, APTHORP & CO., 110 Broadway, N. Y.**

This magazine advertisement for Llewellyn Park appeared in *Appleton's Journal* on December 17, 1870. The most notable landmark of Llewellyn Park is the gatehouse. Very few people today realize that in this old building survives both the forgotten spirit of Haskell's vision of the future and an exact replica of his house that once stood at Eagle Rock.

Shown here is the cliff at Eagle Rock as it appeared c. 1894. Compared to today, the mountainside was somewhat barren. However, riding along present-day Mountain Avenue in the fall when the trees are bare, one can still distinguish the landmark. Eagle Rock sits 660 feet above sea level, and from this vantage point is a spectacular panoramic view spanning the entire valley east to the Hudson River with a skyline view of New York City, Newark, and south to Newark Bay. Of course, over the years the landmarks have changed but the view has not. At the time of this photograph, the tallest structure in New York was the Brooklyn Bridge, which cannot be viewed from Eagle Rock very well if at all now. At the beginning of the 20th century, Eagle Rock began to increase in popularity and offered visitors a view of a constantly changing skyline. Through all these other changes, Eagle Rock has remained largely the same as it enters the 21st century.

Two

THE RESORT
ON THE MOUNTAIN

EAGLE ROCK,

Orange Mountains,

New Jersey.

PRICE 15 CENTS.

Most visitors to Eagle Rock today fail to realize that the empty, open field was once a resort area. This field, known as the Highlawn, contained at least six buildings as early as 1878. This cover came from a *c.* 1895 pamphlet that was several pages and served as a guide to the activities and accommodations available at Eagle Rock.

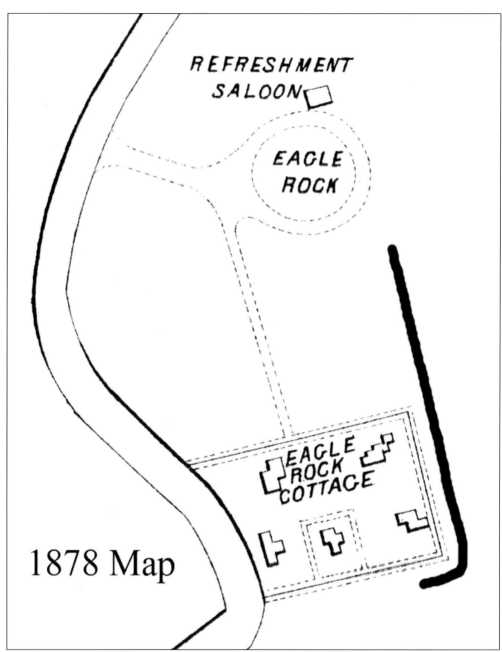

REFRESHMENT
SALOON

EAGLE
ROCK

EAGLE
ROCK
COTTAGE

1878 Map

By the publication of this 1878 map, Eagle Rock was well on its way to becoming a popular weekend resort. The heavy black line to the right represents where the concrete wall is today. The 450-foot wall was not constructed until 1909. At the time of the map, only a small wooden fence ran along the edge and provided some form of protection from the danger of falling down the steep hillside and cliff. The other buildings included a cafe, a hotel, a cottage, and an inn. The refreshment stand was located very close to the actual cliff known as Eagle Rock. This building was torn down c. 1909, when construction started on a building known as the Casino in approximately the same location. None of the buildings shown here stand today. Most were gone by the early 1900s; the last one to survive was the cafe, which was razed c. 1910.

One of the first buildings at the entrance to Eagle Rock Park in the 1870s was the cafe. It also served at one time as a beer garden and an ice-cream pavilion. This photograph, taken c. 1890, shows a local delivery by a horse-drawn wagon most likely from one of the many nearby farms or dairies.

In this view, looking north from the cafe, is the building shown as the cottage on the map to the left. This two-story Victorian house offered guests a quiet, relaxing atmosphere in stark contrast to city life. It really was a breath of fresh air, as most of the bigger cities at that time experienced air pollution from the many local industries.

The cottage, shown here c. 1890, was the center of much activity. Guests strolling the grounds were forever reminded of the inherent natural beauty at Eagle Rock. Behind the cottage and to the right is a horse-drawn buggy at the beginning of a small dirt road leading to a zoological garden.

This road led north from behind the cottage. The building to the left bears a "Photographs" sign on the side and was perhaps some sort of photographic studio. The zoological garden appears to the right. This garden perhaps functioned as a petting zoo or a showcase of the small game that inhabited the surrounding woodlands of Eagle Rock.

Very close to the edge of the overlook was another Victorian house perhaps used as a hotel. Today, no evidence exists of this grand structure or any of the others once gracing the scenic overlook at Eagle Rock. The buildings last appeared on a 1907 map.

THE EAGLE ROCK HOTEL COMPANY, West Orange, N. J.

The Eagle Rock Hotel Company, who now control the Rock and forty acres of Lawn and Grove adjoining, have expended a large amount of money in beautifying this famous resort. Everything for a day or more pleasure may be found on the grounds, which are lighted by electricity as well as properly policed. The view from the grounds at night is grand beyond description; a splendid place for a days ramble. Rooms to be had by the week, month, or season. Cuisine and service of high standard. Livery, Telegraph Office, Mineral Spring, fine roads diverge in all directions leading through wildwood and forest, over mountains and through valleys. Open the entire year. Telephone 483 Orange.

This c. 1890 advertisement for the Eagle Rock Hotel Company boasts of a telegraph office and electrical lighting at Eagle Rock. Rooms could be rented for an entire season, suggesting that those of financial means could spend the entire summer enjoying the cool mountain breezes.

This *c.* 1895 photograph shows British-born Hall of Fame boxer Robert Fitzsimmons (1863–1917), who once trained at Eagle Rock. He held three world titles: middleweight (1881–1897), heavyweight (1897–1899), and light heavyweight (1903–1905). He most likely trained at Eagle Rock in 1894, when he fought in Newark on January 26, and again on March 21 of the same year.

Fitzsimmons may have liked Eagle Rock because it was a quiet, secluded place. The many trails and paths must have provided a scenic backdrop for the 10- to 15-mile runs Fitzsimmons performed during his training. This view, from a *c.* 1898 postcard, shows a lonely path near Eagle Rock.

An unidentified young man sits on top of one of the many giant boulders scattered along the hillside below the Eagle Rock cliff. At the time of this photograph in 1897, one could explore the rugged terrain on foot; however, today the area along Mountain Avenue is fenced in and overgrown with many trees and shrubs, making the trek difficult and dangerous.

Located throughout Eagle Rock's many paths and trails were rustic shelters. They provided a peaceful place to rest in the shade or to seek protection from a rainstorm. This c. 1899 postcard view shows a water pump in a shelter. Although located at a higher elevation, the water table on top of the mountain is relatively shallow.

This Victorian structure, part of the Eagle Rock Hotel Company c. 1895, was most likely used as a hotel and sat across from a wide dirt road where a wooden fence ran along the edge of the cliff. Peeking between the trees in the lower center is the porch of an adjacent guesthouse or inn.

A closer look at the old wooden fence along the edge of the cliff, c. 1900, shows how frail it must have been. The sign to the right reads, "Massacre Rock," perhaps a lighthearted warning of the danger of leaning too close to the cliff. In just a few short years from the time of this photograph, a concrete wall replaced the crumbling fence.

Guests pose c. 1895 on the front porch of a house that stood on the winding road around the front of the overlook on the Highlawn. When the Essex County Park Commission was formed in 1895 and land purchases began at Eagle Rock, these quaint structures began to disappear.

In this c. 1907 photograph depicting the same building as above, the porch has been enclosed and the upper balcony removed. The Eagle Rock Inn was probably the last surviving link to the resort that once existed at Eagle Rock before the Essex County Park Commission takeover.

Shown here c. 1910, the cafe was the last structure standing from the Eagle Rock resort days of the late 1800s. As the Essex County Park Commission acquired the land at Eagle Rock, many changes and improvements were made. The annual report of 1900 stated that most of the buildings had been removed and the land cleared.

This building served originally as a cafe and restaurant and then in later years as a beer garden. In its last years, it was an ice-cream pavilion. It was torn down c. 1911, an act that cleared all structures from the Highlawn and created the open field of Eagle Rock that is recognized today.

Young ladies enjoy a leisurely afternoon in the park *c.* 1910. At the far left, a woman and child make their way to the ice-cream pavilion. Just to the left of the tree and under the first branch the Eyrie appears. It was the first house built at Eagle Rock in 1857 and the last one torn down in 1924.

Most of the shelters scattered throughout the woodlands of Eagle Rock were open-air structures. This shelter, pictured *c.* 1895, however, had a roof and what appears to be a chimney. Perhaps it was a used as a camping cabin for seasonal visitors to the Eagle Rock resort. The shelter's location is unknown.

This building is shown on the 1878 map on page 20 as the refreshment saloon. Pictured here in 1895, it included an open-air area where visitors could enjoy a cold lemonade away from the hot sun. The building was torn down prior to 1909, when construction started on the Casino in about the same location. Today, the Highlawn Pavilion restaurant occupies this site.

As Mountain Avenue crosses into Montclair from West Orange, it changes into Undercliff Road. The road was so named for the cliff it ran under: Eagle Rock. In this rustic 1895 setting, a manhole cover appears in the foreground as if staking claim to the future development that would occur around Eagle Rock.

The Essex County Park Commission was formed in 1895 and began to purchase the land at Eagle Rock to become part of the Essex County Park System, believed to be the first in the United States. The first president of the newly formed commission was Cyrus Peck, and the first vice president was Frederick M. Shepard. Pictured here is Howard Hayes, who was a member of the commission in 1903.

Eugene Vanderpool served the Essex County Park Commission from April 17, 1899, to July 12, 1903. During this time, less then 18 acres was added to Eagle Rock land purchases. By 1907, the entire tract of 410 acres comprising the Eagle Rock Reservation was completed, for a total cost to Essex County taxpayers of $256,975.59.

Pictured here is Robert F. Ballantine. He was appointed to the Essex County Park Commission on December 24, 1901. He was intended to finish the term of Franklin Murphy, who had left for the statehouse in Trenton when he was elected governor. Ballantine, however, died before that term was completed, passing away on December 10, 1905.

Franklin Murphy gained fame for his outstanding record in the Civil War. He served on the first park commission and left when elected governor of New Jersey in 1902. Following his term of office, he later served as the commission's vice president and finally as president from 1911 to 1920. He was dedicated to public service but also founded a successful varnish business in Newark.

Three

THE EAGLE ROCK TROLLEY

By the 1890s, Eagle Rock had gained such popularity that it became necessary to provide a means of cheap transportation to the mountain resort. On June 20, 1894, a trolley line to the foot of Eagle Rock on Mountain Avenue was placed in operation. This *c.* 1910 photograph shows Car No. 111 descending the mountain at the present-day intersection of Nutwold Avenue and Moore Terrace.

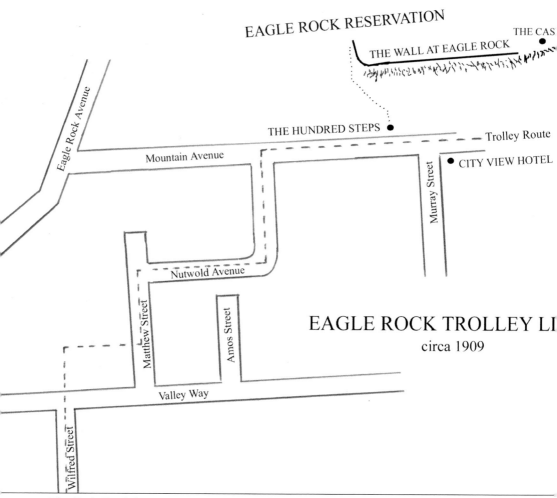

EAGLE ROCK RESERVATION

THE CAS

THE WALL AT EAGLE ROCK

THE HUNDRED STEPS ● — — Trolley Route

Eagle Rock Avenue

Mountain Avenue

CITY VIEW HOTEL

Murray Street

Nutwold Avenue

Matthew Street

Amos Street

EAGLE ROCK TROLLEY LI

circa 1909

Valley Way

Wilfred Street

The Eagle Rock line was an extension of the Washington Street end of the Orange Crosstown Line. It ran via Washington Street to its own right-of-way along what are now Watchung Avenue, Chestnut Street, Oxford Place, and Cherry Street, to Harrison Avenue and via that street to its intersection with Eagle Rock Avenue. It then proceeded along Mississippi Avenue, up Wilfred Street crossing Valley Way, and turning right to Matthew Street, which is currently Moore Terrace. It ran up Moore Terrace, again making a right turn up what is today Nutwold Avenue, reaching Mountain Avenue and turning right on its zigzag path up the mountain. Continuing along Mountain Avenue, it terminated at the City View Hotel at the foot of Eagle Rock on the West Orange–Montclair border. It would have been impractical, if not impossible, to run a trolley directly to the top of Eagle Rock because of the excessive grades.

This photograph shows the car house on Washington Street in Orange c. 1903, where service to Eagle Rock originated. The trolley venture was not a financial success, and in 1898, the Suburban Traction Company, which owned the Eagle Rock line, went into receivership. It was taken over by the Orange and Passaic Valley Railway Company. In 1903, the company became part of the Public Service Railway System.

This 1914 ticket transfer for the Crosstown Line shows service available from the car house to the Eagle Rock line. During peak summer travel to Eagle Rock, the cars were so crowed with passengers standing inside and clinging to the running boards outside that the conductor had to be an acrobat and an optimist if he expected to collect all the fares.

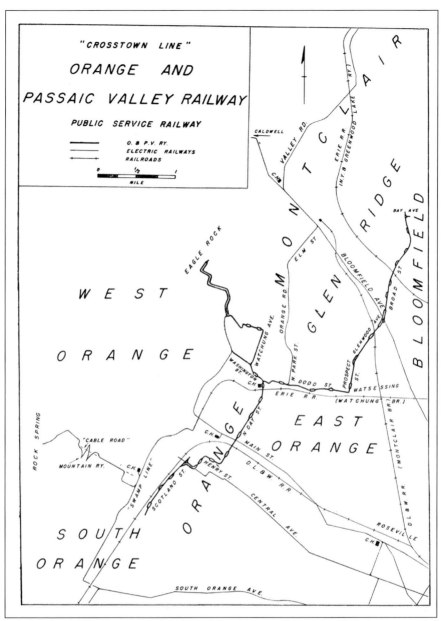

Shown here is a system map for the Crosstown Line, operated by the Orange and Passaic Valley Railway. The Eagle Rock line was a branch of this system. This *c.* 1901 map reveals the various electric street railways and railroads operating in and around the West Orange–Orange area at the time. In the days before the automobile, Eagle Rock was accessible from all points through this vast network of connecting systems. Those seeking refuge from the congestion of city life in such places as Newark could experience a refreshing outing to the rural setting of Eagle Rock and return the same day. At this time, a small ice pond just to the west of Eagle Rock at the intersection of Eagle Rock and Prospect Avenues attracted attention. The owners rented rowboats to picnickers. The spot soon became equally as popular as Eagle Rock, and a restaurant, dance hall, and merry-go-round were constructed. The enterprise was called the Crystal Lake Amusement Park, now known simply as Crystal Lake.

Eagle Rock Avenue, as shown in this c. 1904 postcard, was nothing more than a dirt path where pedestrians easily outnumbered the occasional automobile. Once visitors arrived at Eagle Rock by trolley, they needed only take a short stroll down Eagle Rock Avenue to Crystal Lake. As the buildings at Eagle Rock started to disappear after 1900, the small lake gained in popularity.

Although separate places and only a short distance from one another, both Eagle Rock and Crystal Lake became favorite destinations largely due to the trolley line. The previous view looks west on Eagle Rock Avenue just after leaving Eagle Rock, and this view, also c. 1904, points east back toward Eagle Rock from near the intersection of Prospect and Eagle Rock Avenues.

At the end of the Eagle Rock line on Mountain Avenue was the City View Hotel. From this point, just across the street, visitors had to climb a wooden zigzag staircase the last few hundred feet to the top of Eagle Rock. This c. 1905 postcard view shows that this could be a place full of activity.

The City View Hotel was also known as Cox's Hotel, presumably because the proprietor was John Cox. George Underwood was also once a proprietor here. The hotel was located at the present-day corner of Mountain Avenue and Murray Street. Ice-cream sales here were a welcome treat on a hot summer day. Eagle Rock Car No. 51 prepares to depart, heading to Orange, c. 1900.

This *c.* 1904 postcard view shows a wintry scene along Mountain Avenue. The City View Hotel appears down to the right. The tracks of the Eagle Rock trolley are visible through the light snow. Service to Eagle Rock was not seasonal but was certainly more popular in the spring, summer, and early fall.

Public Service Car No. 6 is pictured at the Plank Road Shops in Newark *c.* 1907. The car was a former Orange and Passaic Valley Railway car used on the Eagle Rock line. It was 16 feet long and was equipped with two General Electric model 800 motors that were used for traction in negotiating the steep grades up to Eagle Rock.

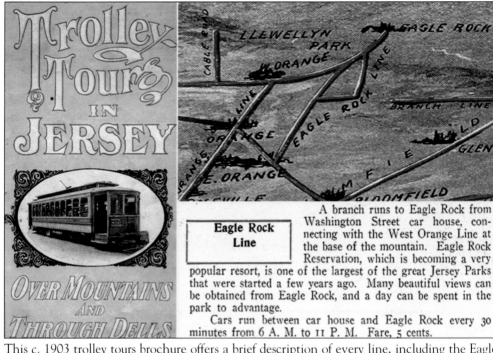

Trolley Tours IN JERSEY

OVER MOUNTAINS AND THROUGH DELLS

Eagle Rock Line

A branch runs to Eagle Rock from Washington Street car house, connecting with the West Orange Line at the base of the mountain. Eagle Rock Reservation, which is becoming a very popular resort, is one of the largest of the great Jersey Parks that were started a few years ago. Many beautiful views can be obtained from Eagle Rock, and a day can be spent in the park to advantage.

Cars run between car house and Eagle Rock every 30 minutes from 6 A. M. to 11 P. M. Fare, 5 cents.

This *c.* 1903 trolley tours brochure offers a brief description of every line, including the Eagle Rock line. It advertises the Eagle Rock Reservation as one of New Jersey's great large parks. On the system map, Eagle Rock appears as a remote outpost. According to the brochure, a nickel got you there.

Pictured in this *c.* 1921 Dyer family photograph is Dominick Dyer (1886–1953). He worked as a motorman on the Eagle Rock trolley line and in later years as a chauffeur for Mrs. John E. Sloane, Thomas Edison's daughter. His friends and family lightheartedly referred to him as "Dominick Dyer, the Eagle Rock Flyer."

Public Service Car No. 889 is shown at the Plank Road Shops in Newark c. 1907. This trolley was former Orange and Passaic Valley Railway Car No. 63, which was also used on the Eagle Rock line. Before World War I, this was one of two open-air cars used. Built in 1902 by the Laclede Car Company, it was equipped with 10 benches for passengers.

This c. 1915 postcard scene depicts a lonely Eagle Rock in winter boldly looking out over the barren hillside along Mountain Avenue. The tracks of the Eagle Rock trolley are visible on the street as if paying silent homage to Eagle Rock. In less then 10 years, the trolley was gone, replaced by the emerging technology of the day—the automobile.

Suburban Traction Car No. 101 appears on the Eagle Rock line at the bottom of Nutwold Avenue c. 1895. The trolley prepares to round the curve and ascend the steep grade up to Mountain Avenue en route to the City View Hotel terminus. This was one of the two open-air cars used on the line.

When the trolley tracks were torn up in the late 1920s, the rails were not scrapped but used as posts of a guardrail at the bottom of Nutwold Avenue. As seen in the highlighted portion of this contemporary photograph, the old rails, despite some rust, still survive today as a reminder to a forgotten past.

A car has just left the City View Hotel *c.* 1904 and is headed down Mountain Avenue, about to turn onto Nutwold Avenue. On the evening of June 5, 1903, Car No. 51 (seen on page 38) derailed on this curve. Fifteen passengers were injured, some as a result of jumping from the car. A prankster was believed to have released the brakes, causing the accident.

A horse-drawn buggy rumbles down Mountain Avenue past the City View Hotel (just out of view to the left) *c.* 1905. The trolley makes the turn at the top of Nutwold Avenue onto Mountain Avenue in the distance. To the right are the Hundred Steps, which lead to the top of Eagle Rock.

A trolley bound for Orange fights its way through a maze of cars and people at the bottom of Eagle Rock Avenue and Main Street in West Orange c. 1901. The Eagle Rock trolley also passed close to this intersection. As automobiles increased in numbers, ridership on small branch lines like Eagle Rock began to decrease. By the 1920s, trolley use all over the United States was on the decline.

This photograph was taken on the morning of April 19, 1924, the last day of operations for the 30-year-old Eagle Rock trolley. Shown here, from left to right, are James Costigan (a motorman for 25 years), William Pierce, and unidentified. No one now will ever know the joy of riding the Eagle Rock trolley, but it is nice to know somebody once did.

Four

THE HUNDRED STEPS

Since it was impossible for the trolley to make the steep grade to Eagle Rock, visitors climbed a zigzag set of wooden steps to reach the top. This staircase, known as the Hundred Steps, was located on Mountain Avenue just across the street on a slight angle from the City View Hotel at the end of the Eagle Rock trolley line.

After just a few small steps from the road, visitors turned left to traverse the mountain grade on an angle. Whether or not there were actually 100 steps to climb is not known. In this *c.* 1906 photograph, above in the distance is a turn back to the right.

This card, dated September 20, 1907, gives a closer view of the right turn in the above photograph. The postcard reads, "Dear Regina, Received your postal, and was so glad to hear from you. Am having a very good time. Was aboard the Lusitania yesterday. Love from Gertrude." Most likely, the sender was a visitor aboard the *Lusitania* while it was moored in New York Harbor after its maiden voyage.

The Hundred Steps emerge at the top of Eagle Rock alongside the wall in this c. 1916 postcard. This concrete retaining wall was constructed in 1907 to replace the old wooden fence, which had become unsafe. A leisurely stroll at one's own pace, combined with the natural beauty along the way, compensated for the difficulty of having to make the grade by foot.

Another pathway at the beginning of the Hundred Steps on Mountain Avenue veered to the right. This trail combined walkways and steps and also went to the top of Eagle Rock. It, however, was not the most direct route. It wound along the hillside under the cliff's edge, offering those who dared a closer look at the rock formations on the face of Eagle Rock.

This barren landscape offers an unobstructed view of the Hundred Steps in late fall or winter in this c. 1905 postcard. The staircase has been neatly cut into the steep grade of the hillside. The tracks of the Eagle Rock trolley lie in the foreground along Mountain Avenue.

This c. 1911 postcard reads, "Path to Eagle Rock, Top of Orange Mountain, NJ." Eagle Rock is located in West Orange and partially in neighboring Montclair and Verona. Period postcards often refer to Orange Mountain, which can lead one to believe that Eagle Rock is in Orange. The Orange Mountains are part of the Watchung Mountain Range in New Jersey, and nearby Orange is in the valley.

Half the fun of visiting Eagle Rock was getting there. After an enjoyable and sometimes crowded trolley ride and a vigorous climb up the Hundred Steps, gazing from the overlook offered an opportunity for some much-needed rest. On a clear day, one could spot his own house from Eagle Rock. This location today is easily recognizable and has not changed much since this c. 1901 postcard view.

The Hundred Steps certainly did not provide the only access to the top of Eagle Rock but were just convenient for those using the trolley. Those who decided to walk usually came straight up Eagle Rock Avenue, shown here c. 1905. In the days before 1951, the road led directly into the park. Today, this abandoned stretch of old Eagle Rock Avenue remains in the park overgrown with brush.

After the trolley service was discontinued in 1924, the Hundred Steps had outlived their usefulness. They remained on the hillside for years after but slowly declined into a state of disrepair. Years exposed to the weather without proper maintenance caused them to rot and fade from memory.

This view shows the Hundred Steps c. 1903. Today, this site along Mountain Avenue in West Orange lacks any evidence that the steps once existed. Only the grade cut into the mountainside can be seen with the trained eye.

Five

THE CASINO

In the early part of the 20th century, the Essex County Park Commission started to make improvements at Eagle Rock. The commission was determined to take advantage of the striking natural features of the mountain. In 1909, construction of a building known as the Casino, pictured here *c.* 1914, was begun near the edge of the cliff.

The Casino was located near the edge of the cliff so that visitors did not miss the effect of looking directly out across the valley below. It was situated approximately 1,000 feet from the entrance on Eagle Rock Avenue. The intention was to protect the view from Eagle Rock itself and to keep an area open for horses and carriages to congregate.

The Casino was built 104 feet long and 25 feet wide with a 50-foot rear extension. The two-story building was constructed of brick with a stucco finish and a tile roof. The first floor was used as an open shelter. A dining room—with balconies—was located on the second floor. The extension contained the living quarters of the attendant. The Casino opened in 1912.

The term "casino" is misleading by today's standards. No gambling ever took place here. In the early days of the 20th century, "casino" meant an open-air structure or dance hall. With all the changes taking place at Eagle Rock at the time, this building quickly became a landmark and a popular attraction.

As beautiful as this building was, it was plagued with constant problems. As early as 1914, it was reported to be in a state of disrepair. In 1917, the Casino was turned over to the federal government for research purposes. Thomas Edison performed experiments for the U.S. Navy on the second floor during World War I (see chapter 7).

By the 1950s, the building still presented a respectable outer appearance, disguising the poor nature of the inside. By the early 1960s, or perhaps earlier, the building had closed, and only a small refreshment stand was operated out of a side window by a private vendor on weekends.

A report by the secretary to the Essex County Park Commission in 1970 reported that the building was in a complete state of disrepair. Once the maintenance had stopped years earlier, the elements of vandalism and weather took their toll on the structure. The once popular Casino, abandoned by time, stood as only a remnant of the past.

Pictured here is the Casino, as seen from the other side, c. 1980. Broken windows, crumbling walls, a leaking roof, and deteriorated floors were only some of the problems. No money in the county budget was allotted for the necessary repairs, which meant for an uncertain future, much like the similar situation involving Haskell's home, the Eyrie, 60 years before (see chapter 1).

Even in disrepair, the deteriorating building displays signs of its once magnificent beauty and unique architectural style. No one seemed to notice or appreciate it as it stood as a decaying monument to a bygone era. Sadly, it was described as an eyesore in an otherwise beautiful park.

Sunlight glistens through the open arches of the dark and abandoned Casino. Although in dire straits, the building is not quite ready to depart from Eagle Rock, desperately clinging to the belief that the lively sounds of a generation past will once again echo through its halls.

To the right, the crumbling stucco reveals the brickwork supports of the open arches. By the mid-1980s, Essex County officials were faced with the harsh reality of having to tear the building down. This picture, however, symbolically captures the proverbial light shining at the end of the tunnel down the dark corridor, because help is on the way.

56

Fortunately, a solution was found that spared the graffiti-ridden, deteriorating building. A contract with a restaurant would be granted if the building was renovated at private expense. The Knowles family, who already owned and operated a nearby restaurant in West Orange, accepted the challenge. The details of the contract were negotiated, and the much-needed renovations began in 1985.

On December 19, 1986, Eagle Rock celebrated a new beginning for the old Casino, which had originally opened in 1912. The Highlawn Pavilion restaurant became Eagle Rock's new landmark. Thanks to the vision of the Knowles family and its commitment to excellence, future generations will know this building as it enters the 21st century with a new identity and a treasured past.

Shown here is Eagle Rock on Easter Sunday in 1943. PO Pat Brennan dances with Ann Costigan (back turned) while PO Ken Tinquist, Eleanor Petruzzi, and Joe Fitten pose. The Casino appears in the background during better days. Servicemen home on leave often visited Eagle Rock. Tinquist and Petruzzi married before the war's end, and coincidentally, Ann Costigan's father was a motorman on the Eagle Rock trolley (lower photograph on page 44). It would have been a monumental mistake for Essex County to tear this building down some 40 years later when it fell into disrepair. The building had become such a landmark at Eagle Rock that to lose it could have very well meant the beginning of the end for Eagle Rock in both spirit and character. Sparing it had once again given the building a heart and soul that survives in the Highlawn Pavilion.

Six

THE AUTOMOBILE COMES OF AGE

It is unknown when the first automobile arrived at Eagle Rock, although it must have been a cumbersome sight seeing that first horseless carriage rumbling down the road. Just getting to the top of Eagle Rock in the first crude automobiles would have been an accomplishment. No one could have dreamed of the new era that these machines soon ushered in.

With the advent of improved roads and automobiles, Eagle Rock soon became a popular destination for motorists. In 1909, the Essex County Park Commission constructed the roadway Crest Drive from Undercliff Road to the summit of the mountain so that the Casino and Eagle Rock itself would be more accessible. This picture was taken in 1912. (Courtesy of the Montclair Public Library.)

As automobile traffic increased, trolley ridership decreased, and by 1924, service on the Eagle Rock trolley was discontinued. By the late 1920s, automobiles crowded the drive along the wall nearly every weekend. This trend never stopped and continues to this day.

Also in 1909, the Gates Avenue entrance was added. This improvement provided a way to reach Eagle Rock by means of a winding roadway up the mountainside starting from Undercliff Road. The new roadway, finished with a tar macadam surface, cost $5,875 to construct.

Local residents often refer to the winding road by the unofficial name of Snake Road for the many turns that resemble a snake. These bends make it easier to achieve the steep grade up to Crest Drive. Snake Road is pictured here before 1909 when it was not paved.

By the 1930s, automobiles were well on their way to displacing trolleys throughout America. Visitors to Eagle Rock came by the carload, some to spend the entire day picnicking and relaxing in the park. Others just drove through to catch a quick glimpse of the changing landscape over the cliff's edge, also brought on, in part, by the urban sprawl caused by the automobile.

Over the years, as the steady stream of automobiles and people to Eagle Rock continued, only the style of the cars and dress changed. In this c. 1940s photograph, a group of friends huddles at the wall. When the wall was constructed in 1907, iron spikes were added on the top along its length to prevent accidents. (Courtesy of the Montclair Public Library.)

Pictured here in June 1941, West Orange residents Philip Schoen (left) and Harry McFayden have just returned home from Virginia Beach, Virginia. The group of four young men on the trip also included James Fagan and Roy Johnson (not pictured). Upon returning home, they first stopped at Eagle Rock. This 1929 Ford Model A had a rumble seat and side luggage racks, which, as shown, the group used well. Everyone could not ride inside the car at the same time, so they each had to take a turn in the outside rumble seat. The bald tires perhaps best symbolize the carefree attitude of the young men on their journey. Both men pictured later served in the U.S. Navy during World War II—Schoen in the Pacific and McFadyen aboard the battleship USS *Massachusetts*. Both returned home after the war, perhaps again first stopping to visit Eagle Rock.

The main road leading to Eagle Rock is Eagle Rock Avenue, one of the very first roads in the area. It takes advantage of a natural notch in the rock at the top of the mountain, allowing easy passage. The steep grade made it impractical for the trolley to follow this route. Eagle Rock Avenue is shown here *c.* 1907.

In November 1901, the New Jersey Automobile Club sponsored, in what would become an annual event for many years, the Eagle Rock Hill Climb, a race to the top of Eagle Rock up this steep grade. In those days, Eagle Rock Avenue was nothing more than a well-traveled dirt road.

In the first Eagle Rock Hill Climb in 1901, a car races down and gains momentum at the end of Harrison Avenue, near the top of Cherry Street. It will then begin its climb up Eagle Rock Avenue. Ironically, the car is just about to cross the Eagle Rock trolley tracks, which were abandoned in 1924 largely due to the automobile.

This picture from the same year shows the crowds gathered to watch the races at the present-day intersection of Main Street and Eagle Rock Avenue in West Orange. Making it to the top of Eagle Rock was an accomplishment, but making the steep grade in high gear brought added bragging rights. In November 1951, a similar race was run on the 50th anniversary. (Courtesy of the Newark Public Library.)

This *c*. 1930s view shows the old route of Eagle Rock Avenue, which ran directly into the park and then made a left turn, continuing toward Prospect Avenue. In 1951, Eagle Rock Avenue was realigned to make easier gradual turns, and a new entrance to Eagle Rock was created slightly farther west.

This *c*. 1970s photograph again reveals how car styles have changed since the automobile came of age at Eagle Rock. The overlook appears in the background, and one of the many park benches scattered about at Eagle Rock is at the right. These benches, which date back to the early 1900s, are no longer used.

Seven

THOMAS EDISON
AT EAGLE ROCK

Thomas Edison, pictured here *c.* 1905 in his West Orange laboratories, performed secret experiments for the U.S. Navy during World War I on the enclosed second floor of the building at Eagle Rock. He also did some other experiments in and around the hills of West Orange, including some right on the open fields at Eagle Rock.

The Essex County Park Commission have placed at my disposal a concrete building not used in winter x This is on top of Eagle rock, —— ft above the sea, from which an unobstructed area of 40 miles can be observed, x I shall there make further Experiments on increasing the range of vision, Especially in low visibility due to Haze, fogs & darkness, all of which I will report later— Yours &c

In 1917, during World War I, the Essex County Park Commission reported in its annual report that it had turned the Casino over to the government for experimental purposes and provided a police guard to ensure privacy and protection. Shown here is a handwritten laboratory note by Thomas Edison, with his famous scribbled initials signature. It reads, "The Essex County Park Commission have placed at my disposal a concrete building not used in winter. This is on top of Eagle Rock ——ft. above the sea, from which unobstructed area of 40 miles can be observed. I shall make further experiments on increasing the range of vision, especially in low visibility due to haze, fog and darkness. All of which I will report later." Most of Edison's experiments in this building still remained classified until just a few years ago. The ones mentioned above could be experiments with sonar and a submarine-detection device.

Having been appointed a member of the Naval Consulting Board

I, *Thos A Edison* , do solemnly swear (or af-
        (name in full.)
firm) that I will support and defend the Constitution of the
United States against all enemies, foreign and domestic;
that I will bear true faith and allegiance to the same; that
I take this obligation freely, without any mental reservation
or purpose of evasion; and that I will well and faithfully
discharge the duties of the office on which I am about to enter: So help me God.

Thos A Edison

New Jersey
(State of.)
                    : ss:
Essex
(County of.)

Sworn to and subscribed before me this ___19th___ day
of *September* , 19*16*.

F. S. Curtis
Chief Clerk,
Navy Dept.

This oath of office was signed by Thomas Edison in conjunction with his work and experiments for the navy during World War I. He was appointed to the Naval Consulting Board and served as its chairman. The board consisted of 23 members of various backgrounds and experience from the private sector. Hudson Maxim, also from New Jersey, a proponent for a strong national defense, was a member. The board worked partly in an advisory capacity to the navy on the research and development of new weapons. Of special concern was the need to protect allied vessels crossing the North Atlantic that were being torpedoed and sunk by German U-boats. U-boats posed a constant threat since they were largely undetectable. Some of Edison's experiments focused on the ability to detect submarines at sea and weapon systems that could be installed on merchant ships for protection from the ever-present menace.

The device seen here remains unclear, but the location at Eagle Rock is unquestionable and easily recognizable even today. One of Thomas Edison's assistants is connected to an airplane-detection device on February 19, 1917. The circled images are additional ones, perhaps connected as part of a network. Thomas Edison's stamp, "TAE," appears in the upper

left, meaning this was an Edison Company photograph. The view looks somewhat southwest across the open field at Eagle Rock from the Casino, made available to Edison at the time of the photograph.

This photograph, taken on March 31, 1917, bears a notation on the back that indicates the odd device is a range finder. It was perhaps part of the airplane-detection system Edison was working on at Eagle Rock only a month and a half before. It could also have been one of the experiments Edison referred to in his laboratory note on page 68.

This picture, also labeled as depicting an airplane-detection device, has an unclear location, but the spot is likely in Eagle Rock, near the present-day upper picnic area. The road seen in the background is possibly Eagle Rock Avenue. A wood-rail fence close in similarity to the one shown did exist there.

The so-called airplane-detection device is seen connected to listening apparatus mounted on the tower. The wires leading from the tower could be connected in series to other towers, perhaps as part of an experimental early-warning system for approaching airplanes. This view also appears to be at Eagle Rock, although no exact location is indicated.

A closer look at the device shows its simplistic design. The two sound-receiving horns seem to be wrapped with a wool insulation, perhaps to increase the ability to detect certain sound frequencies. The elevation and unobstructed view from Eagle Rock, as stated by Edison, may have been necessary for the development of these experiments.

Joseph Daniels –
Report N⁰ ——

I herewith enclose photographs
of the smoke produced from
Exploding 1 pound of TNT
and one litre of Oleum 20%
Since these Experiments I find
that 1 lb of 20% Dynamite acts
nearly as well –

A Cheap shell can be made for
3 & 4 inch rifles now used on
merchant ships and used in
addition to regular projectile –
at any time it is thought
desirable to hide the ship
from the submarine, or Choke
them in herein –
The fine Oleum in the cloud
has a terrible effect on the
lungs. Oleum 20% of 24 per ton

Edison

74

From the Laboratory
of
Thomas A. Edison,
Orange, N.J. July 26, 1917.

(copy)

n. Josephus Daniels,
   The Secretary of the Navy,
      Washington, D. C.

dear Mr. Daniels: Report No. 41:

      I enclose herewith six photographs
the smoke produced from exploding one pound of T. N. T.
d one litre of Oleum 20%. I have found, since experi-
nting, that one pound o_ 20% dynamite acts nearly as
ll.

      A cheap shell can be made for 3 and 4 inch
fles now used on merchant ships and used in addition
the regular projectile, at any time it is thought
sirable to hide the ship from the Submarine, or choke
e men thereon. The fine Oleum in the cloud has a
rrible effect on the lungs, the price of which is
4.00 per ton for 20% Oleum.

      Yours very truly,

      (signed) Thos. A. Edison.

Compared here are the rough draft and final letter sent as Report No. 41 to the then secretary of the navy, Josephus Daniels, in Washington, D.C. The rough draft on the left is written in Edison's own hand. Because these experiments dealt with explosives, Eagle Rock, although a rural setting, did not provide adequate isolation from the surrounding population to allow for detonation. The actual test explosions took place farther to the south in West Orange on the mountain overlooking the Orange Valley in the abandoned cable car cut. Perhaps only the mixing of the compounds or the assembly of the experimental shells was carried out at Eagle Rock. Edison received a reply back from Secretary Daniels, stating that the navy was interested in these shells, and further testing, based on Report No. 41, would be carried out at the naval proving grounds.

At the Naval Consulting Board's first meeting in Washington, D.C., Thomas Edison (center) acted as the chairman. The committee was originally called the Civilian Advisory Board on Inventions, but the name was changed when it was determined that the examination of inventions would be only a small part of the duties of the board. The main purpose was to cooperate with the government in providing technical knowledge and experience acquired in the prosperous private industries. The efficient mobilization of the industries of the country in aiding the government in a time of war for the production of munitions and weapons was recognized as being of supreme importance. Edison's work and experiments at Eagle Rock and in the hills surrounding West Orange were under the jurisdiction of this board and were for the purpose of national defense. Hudson Maxim, also from New Jersey, is the second person to the right from Edison, with his face turned.

Shown here is the abandoned Casino building c. 1970. The second floor to the right is where Thomas Edison conducted his experiments. As they were secret, only speculation existed over the years that Edison worked here. The Edison-related documents and pictures in this book come from a private collection.

The sunlight streams through the broken pains of glass on the darkened second floor of the Casino, where Thomas Edison once worked, c. 1970. Lack of maintenance and years of vandalism threatened the building's survival. Saving it meant salvaging an important link to Eagle Rock's history. Those who now dine in the renovated second-floor restaurant can pause in amazement over Edison's presence here nearly a century ago.

Hockey Match on the Ice

©February 24, 1898
Thomas A. Edison

One myth that has long been perpetuated is that one of Edison's early films, *The Great Train Robbery* (1903), included scenes shot in Eagle Rock. This is not true; however, an earlier film, *Hockey Match on the Ice* (1898), was filmed at Crystal Lake, only a short distance from Eagle Rock. Many people incorrectly considered the two places part of the same resort at the time.

This grainy image is taken from Edison's film *Hockey Match on the Ice*, which was probably the first time in history that a hockey game was filmed, and most likely the first time Crystal Lake was seen on film. This occurred on February 24, 1898, in West Orange. *Hockey Match* was a 25-second film from the Edison Manufacturing Company, produced by William Heise.

Eight

EAGLE ROCK: IMAGES IN TIME

The wall at Eagle Rock is shown in this *c.* 1921 photograph. It was constructed in 1907 for a total cost of $5,297, replacing the old dry stone wall and wooden fence that had begun crumbling and had become a safety hazard. The 450-foot-long concrete wall runs along the outer edge of the cliff.

WHAT TO SEE AROUND ESSEX

NO. 4—EAGLE ROCK.

FROM Eagle Rock, perched high on the summit of Orange mountain, one can see the homes and workshops of more people than from any other point in the world.

Eagle Rock is part of the reservation by the name, 408 acres in size, which occupies a section of West Orange and narrow strips of Verona and Montclair. It is easily accessible to automobilists.

The rock itself is nearly 625 feet high, and on clear days the skylines of New York are easily seen, with the Hudson, Hackensack and Passaic rivers narrow strips of blue between. It overlooks twenty-two municipalities, including New York.

At night the view from Eagle Rock is especially impressive, as lights from all the towns twinkle down below. The light on the tower of the Chrysler Building, New York, stands high above the others on the horizon.

The highest point in the Eagle Rock Reservation is about 2,000 feet north of the entrance gate in Montclair, where the elevation is 659.7 feet. Few persons, however, visit this point, most visitors continuing on a few hundred yards to Eagle Rock itself, where there are parking facilities for scores of cars.

Eagle Rock is a part of the Essex County Park System. To reach it, drive up Eagle Rock avenue, at the end of Main street, West Orange, or up Prospect avenue from Bloomfield avenue, Montclair.

◆◆◆

A local newspaper, the *Newark Sunday Call,* described Eagle Rock on October 4, 1931. The article draws attention to the spectacular natural vista at Eagle Rock. The spot has always been popular due to the unique view of the greater New York area. The skyscrapers that now define the skyline of New York had just started to emerge on the Eagle Rock horizon. In the early part of the 20th century, the race for the tallest building in the world started, and the Chrysler Building was the first to top the then highest building, the Eiffel Tower in Paris. In May 1931, the Empire State Building opened, becoming the world's new tallest building. Eagle Rock now became a favorite destination to view the constantly changing skyline. Quick and easy access by automobile also helped attract many visitors on a clear night for a ride in the park and a view of the lights.

80

A young couple playfully poses for the camera in an open-air shelter *c.* 1915. Eagle Rock made use of open-air shelters at various locations scattered throughout the wooded areas. Miles of trails and paths made it possible to walk and enjoy the abundant natural beauty. These rustic structures offered a place to just sit and relax or, if necessary, to seek refuge from the rain.

Two young girls carry their small puppies on the rocks of the Eagle Rock overlook *c.* 1924. The man to the left points out over the cliff, perhaps spotting his house in nearby Montclair. In the days before air travel was popular or practical, the high elevation gave onlookers the simulated effect—with some imagination—of looking down from an airplane.

Pictured *c.* 1928 are, from left to right, Jim Fagan, Buddy Cunningham, and Ken Tinquist. The three West Orange youngsters discovered the joys of the Eagle Rock woodlands at an early age, often hiking up the mountain from their nearby homes to enjoy a day of playful adventure in and around the surrounding hills.

Ken Tinquist and his girlfriend, Eleanor Petruzzi, look over the wall at Eagle Rock *c.* 1942. Tinquist served stateside in the U.S. Navy during World War II in Washington, D.C. On leave from the service, Tinquist often returned to Eagle Rock, a place he knew well as a child. The two married in 1943.

Philip Schoen (left) and Jim Fagan are shown near Eagle Rock in January 1946. Fagan has just returned home from service. Both men served in the navy during World War II. Along with their mutual friend, Tinquist, the two were well familiar with Eagle Rock as young boys growing up.

Some 70 years after first visiting Eagle Rock as kids, Jim Fagan (left) and Ken Tinquist (right) continue to stroll the tree-lined paths. Their friendship has spanned the decades and endures today, never straying far from their childhood roots at Eagle Rock.

Carl J. Kress, a 32-year-old bookbinder from Orange, made frequent visits to Eagle Rock in the 1930s. He was issued a permit from the Essex County Park Commission allowing yodeling in the morning between 8:00 and 8:45 at Eagle Rock. Kress obtained the permit in 1936 because the commission's rules prohibited singing or playing musical instruments in the park at the time.

In November 1914, a severe ice storm hit Eagle Rock. The freezing rain broke numerous branches, and many trees had to be removed. A storm of this character does as much damage as a fire. The ice-coated woodlands make for a beautiful sight as they sparkle in the sunlight, but the effects are serious and long-lasting to the forest.

The tree offers a shady spot out of the sun to enjoy the view over the hillside *c*. 1924. Eagle Rock provided a place to relax and reflect on a lazy Sunday afternoon away from the hectic daily activities of city life. The cool mountain breezes gently blow the fresh country air, giving a refreshing change of pace.

The Casino appears just behind the tree to the left as a group of young adults poses at Eagle Rock in June 1941. As the trees and flowers once again begin to bloom in late spring, the cold days and gray skies of winter slowly become a distant memory.

This view looks back across the open field at Eagle Rock c. 1970. The field's official name is the Highlawn, and it first started appearing on maps as such in the early 1900s. The present-day restaurant at Eagle Rock takes its name from this area. On this field during the 1890s, Eagle Rock profited as a resort and boasted various hotels and inns and a cafe.

This c. 1950 photograph shows the wall and drive at Eagle Rock. During this time, coin-operated binoculars were available along the wall, allowing visitors a detailed look at the various landmarks in and around New York. On a clear day, one could see for miles without them, however. The binoculars were removed in the mid-1970s.

The Casino peeks out of the shadows on an autumn day on the left side of this c. 1969 picture. Visiting Eagle Rock in the fall offered a full spectrum of natural beauty as the trees changed color. Once all the leaves were gone from the trees in late fall, an unobstructed view of some of the nearby homes in Montclair and West Orange was achieved.

Here, through the now bare trees, homes directly below the Eagle Rock cliff on Mountain Avenue in West Orange are revealed. During the fall, such views were possible. The street below is where the Eagle Rock trolley once ran. The City View Hotel also once stood near this location (see chapter 3).

From the wall at Eagle Rock, the sunlight is seen rising over metropolitan New York c. 1968. From Eagle Rock, one can view one of the world's most populated areas. On March 17, 1932, the syndicated column "Ripley's Believe It or Not" claimed that from Eagle Rock it was possible to see the homes of 12 million people.

On this three-and-a-half-acre field in Eagle Rock bordering Prospect Avenue in West Orange, the Essex County Park Commission started a sod nursery in 1939. The site was free from tree growth and could be used to grow sod for other parks in Essex County. In the first sod harvest in November 1941, Vailsburg Park in Newark received 4,500 square feet of new sod from Eagle Rock.

On a clear night, the view from the wall at Eagle Rock is a sea of lights, as can be seen in this c. 1968 photograph. The tall building on the horizon to the right is the Empire State Building. It opened in 1931 and has since been a favorite landmark to view from Eagle Rock at night or day.

In this view looking southeast from Eagle Rock c. 1968, the homes in the foreground are located in nearby West Orange. The buildings in the distance are those of downtown Newark. Beyond that are the towers and the faint outline of the Verrazano Narrows Bridge, at the entrance to New York Harbor. When it first opened in 1964, it was the world's longest suspension bridge.

Shown here is the entrance to Eagle Rock *c.* 1940. At this time, Eagle Rock Avenue came straight into Eagle Rock (the road to the right, between the rocks and the path). At the T intersection, one made a right into Eagle Rock or a left to continue on Eagle Rock Avenue. In 1951, Eagle Rock Avenue was realigned, and the current entrance to Eagle Rock was created.

This *c.* 1915 photograph gives an earlier look at the original entrance to Eagle Rock. In the center stands a building that was on the field at Eagle Rock at the time. It was an open-air shelter that was located just across the drive that ran along the wall.

This postcard view from 1915 shows the field at Eagle Rock and the same open-air shelter seen in the lower image on the previous page. The wall along the edge of the drive is seen to the right. This shelter provided a cool spot to sit out of the hot summer sun and perhaps a place for a much-needed rest after climbing the Hundred Steps.

This postcard view of Eagle Rock Avenue was taken just before the intersection with Prospect Avenue in West Orange c. 1904. Eagle Rock appears to the right, and the houses to the left are located at the entrance to nearby Crystal Lake. Crystal Lake was only a short walk from Eagle Rock. One of these houses still stands on Eagle Rock Avenue, with a huge communication tower standing behind it.

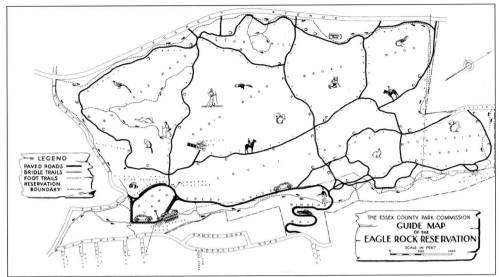

This self-guided tour map was issued by the Essex County Park Commission in 1960. It shows the various trails and paths of the Eagle Rock woodlands. It is predominantly a red oak forest with a wide range of floral growth and with a unique red maple wetland in the northern section of the tract. Eagle Rock truly is an island of forested landscape surrounded by a sea of suburban sprawl.

Twelve fireplaces were built in Eagle Rock's upper picnic area c. 1949. These four-sided fireplaces were removed c. 1975; only picnic tables occupy this area now. To the right is the original open-air shelter, built c. 1910. It was also removed c. 1975, but another one was constructed in its place.

In 1973, Eagle Rock welcomed the newly completed World Trade Center in New York. In this picture, its faint silhouette can be seen peering through the haze on the horizon at an Eagle Rock sunrise. When the twin 110-story buildings were completed that year, they were the new tallest buildings in the world until the Sears Tower surpassed them both in 1974.

An empty park bench at Eagle Rock provides a perfect vantage point to experience the spectacular view of the sea of lights just beyond the cliff's edge. As the World Trade Center slowly began to emerge on the scene, it redefined the New York City skyline. Construction began in 1966 and cost an estimated $1.5 billion.

Shown here is a daytime look at New York City seen from Eagle Rock *c.* 1969. On any clear day, Eagle Rock provides an exceptional and breathtaking view. When the wall was built in 1907 at Eagle Rock, the Flatiron Building, built in 1902, was New York's only skyscraper. Throughout the 20th century, the evolving age of the skyscraper was witnessed from Eagle Rock.

When daily progress starting creeping skyward in 1966, no one actually realized how high 110 stories would be. Viewing the completed towers of the World Trade Center from Eagle Rock, at a distance of about 20 miles, helped put it all into perspective. Here, the sun sets over Eagle Rock *c.* 1974, causing the buildings to briefly glow a brilliant white from the reflecting sun glare.

At the end of the 19th century, no skyscrapers were seen from Eagle Rock. No one had dreamed such a building was even possible. As time passed, however, the continuing progress of the 20th century unfolded across the horizon. The skyline over the wall may have changed, but Eagle Rock itself had not. It remained as a link to yesteryear with a view of a vast metropolis. This picture was taken in 1969, and soon after the World Trade Center emerged as a modern wonder. Sadly, on the morning of September 11, 2001, Eagle Rock witnessed an event that would change the world. As we enter the 21st century, a permanent memorial will be erected on this exact spot for future generations to know the progress and tragedy seen from Eagle Rock.

As the tragic events unfolded on September 11, 2001, Eagle Rock stood as a silent witness while a horrified nation watched. The skyline was forever altered, but for the many lost, Eagle Rock pays a quiet tribute. In the aftermath, many family and friends placed items at the wall in loving memory of all those who perished. It was a dreadful day when our resolve was tested but not defeated, our spirit bruised but not broken. A permanent memorial will forever honor the victims. Eagle Rock will enter the next century standing proudly as a centurion, forever guarding its place in time. For those of us who can see its splendid beauty, the roads of home will always lead through Eagle Rock.